Eureka

JILL BLEE

EXISLE
PUBLISHING

First published 2007

Exisle Publishing Limited
'Moonrising', Narone Creek Road, Wollombi, NSW 2325, Australia
P.O. Box 60-490, Titirangi, Auckland 0642, New Zealand
www.exislepublishing.com

National Library of Australia Cataloguing-in-Publication Data

Blee, Jillian, 1942– .
Eureka.

ISBN 9780908988662.

1. Eureka Stockade (Ballarat, Vic.). 2. Gold miners – Victoria – Ballarat. 3. Riots – Victoria – Ballarat. 4. Gold mines and mining – Victoria – Ballarat – History – 19th century. I. Title.

994.57031

10 9 8 7 6 5 4 3 2 1

Designed by Christabella Designs
Typeset in Bembo 10/13
Printed in China through Colorcraft Limited, Hong Kong

CONTENTS

Chapter 1

GOLD, GOLD, GOLD

Just before dawn on the morning of 3 December 1854, 296 soldiers and police descended on the sleeping inhabitants of a stockade, which had been roughly thrown together by the miners on the Ballarat goldfields. The soldiers had anticipated an attack. The troops gathering at the government camp were spoiling for action, but no one expected it to come on the Sabbath, and the stockaders were caught off-guard. Within fifteen minutes, about 30 men were dead or dying, and the only rebellion to take place on Victorian soil had been quashed.

The issues that drove the miners to erect the stockade were not unique to Ballarat. They had been brewing since the announcement of the first genuine find of gold was made on 16 July 1851, the day after the colony of Victoria was born. At the heart of the matter was the disdain with which the government regarded the gold seekers, even though it had been responsible for encouraging the gold rush in the first place. Worried that the economy of the new colony would suffer if there was a sudden loss of population to New South Wales, where gold had already been found, a group of businessmen, at the instigation of the governor, had put forward a prize of 200 guineas, sparking a frenzy of gold searching across Victoria.

Unfortunately, the government had not anticipated that the first finds would be replicated all over the colony, or that some goldfields would be so rich that almost the entire male population would be tempted to take up their picks and panning dishes, and tramp into the country to seek their fortunes. Nobody wanted to stay working for wages on the pastoral stations or in the towns. In trying to protect the colony, the government had brought it close to ruin, but rather than adapt to the altered situation, it attempted to drive the men back to their jobs by treating them as an unruly rabble.

They were anything but. They were, in fact, quite orderly, particularly in the early days. Among those who joined in the search for the precious metal were professional men who had never before held a pick or shovel. They took their place beside tradesmen, labourers, and shepherds who had the advantage of knowing the Victorian bush better than anyone else. There were also a few ex-convicts from Van Diemen's Land, which came to be known as Tasmania in 1856. Admittedly few, if any, had asked the permission of their employers before heading to the goldfields, but they were hardly lawless. The early goldfields were remarkable for their lack of crime. When disputes arose, usually over the territory each man had claimed for himself, they came together as a group and sorted the matter out. These mass meetings became a feature of the goldfields, much to the concern of the administration because of the way in which they were conducted. They were democratic. Every digger, as all who joined the search for gold were known, had the right to voice his opinion and to vote on the issues which concerned him, a right which the British Government did not believe was an entitlement of men without property or respectability. Democracy, like republicanism, had no place in the British Empire.

Paying for the privilege

According to British Law all precious metals on or below the ground belonged to the Crown; the diggers were not only trespassing, they were stealing as well. In New South Wales, this problem had been overcome by the introduction of a licence, costing 30 shillings, which gave the miners the right to extract gold from a small patch of land 2.45 metres square for 30 days. An identical licence was introduced into Victoria on 1 September 1851, not so much for the revenue it would raise, though that was sorely needed, but in the hope that the cost would dampen the diggers' enthusiasm for goldmining.

By then, news had begun to circulate about Poverty Point, a stretch of land 800 metres long and 200 metres wide at Ballarat. It had once been a bend in a fast flowing river, and it was strewn with gold nuggets. Renamed

Golden Point, it was to yield much more gold once men with Californian experience had arrived to unearth the pipe clay that lay 3 to 12 metres below the surface. It was so rich that over two days in early September, two brothers brought 27 kilograms of gold out of a shallow shaft. The diggers were prepared to suffer any conditions — wind, rain, mud and the flimsiest of shelter at night — to get their hands on similar amounts of gold, especially as the selling price was £3 an ounce, more than a working man could expect to earn in a week.

To see that the licence was taken up, gold commissioners had to be appointed to every major goldfield and they had to be supported by police who could arrest defiant diggers. The problem was most of the police in the colony had taken off to the goldfields themselves so the native police were called into action. Formed in 1842 when Victoria was still the Port Phillip District of New South Wales, their role had been to assist in the pacification of Aborigines in the pastoral districts. They relished the opportunity to exert their authority over white diggers and were given ample encouragement by their commanding officer, Captain Dana. They arrived in Ballarat on 19 September along with Commissioner Francis Doveton.

He brought with him a bundle of freshly printed licences, but not nearly enough for the 5000 diggers he found on the goldfield. Thousands more licences had to be hastily handwritten so they could be introduced immediately, even though he had been instructed to wait until the beginning of the next month. The diggers were immediately upset. They vowed to resist, especially as the possession of a licence gave them no guarantee that their gold would be safely transported to banks in Melbourne or Geelong. Captain Dana's police were too busy to provide an escort, unless it was to protect the gold of the diggers who had paid for their licences.

Few of the diggers wanted to leave the Ballarat field while it continued to yield such an amazing quantity of gold. One by one they capitulated and bought a half licence at 15 shillings to tide them over for the rest of September. Those who persisted in their belief that the licence cost was too high, and an injustice, risked being subjected to some very rough treatment by the native police. As there was no accommodation for prisoners at the government camp, they could find themselves chained to a log until the commissioner or the police magistrate dealt with the charges against them. A fine of £10 was the usual result, although the diggers could be sent to gaol in Melbourne or Geelong, or had to forfeit their claims.

Provoking the diggers

The antagonism between the government camp and the diggers was further inflamed by the announcement early in December 1851 that the licence fee would be double in the New Year. It would also be extended to everyone on the goldfields who made a living either directly or indirectly from gold. This included those enterprising souls who had set up shop in their tents, providing food, grog and everything else the diggers needed.

The rationale for this provocative action was the financial crisis which had resulted from the shortage of labour. The cost of even the most basic of commodities had spiralled out of control, and the government was afraid that the situation would worsen once word of the gold discoveries reached England. There was a fear that it would create such a rush of immigrants that the colony would be overwhelmed, particularly if the gold petered out before the ships carrying them had arrived in Melbourne or Geelong. Even if the gold lasted, food shortages could cripple the colony and possibly lead to famine, which invariably brought disease and death.

Opposition to the increase in the licence fee was immediate across the goldfields and had the potential to become violent, but on Ballarat there were hardly any diggers left to make an impact. Most had left for fresh fields along the Lodden River and at Mount Alexander. Here 'monster' meetings were held with thousands of diggers denouncing the licence as an unjust tax. They were in the process of getting up a deputation to bring their grievances to Melbourne when the increase was revoked, but the damage had been done. The diggers knew they could not trust the government.

The Eureka Lead

Once the initial excitement was over, some diggers drifted back to Ballarat where they found, to their surprise, that a town had been properly laid out by the surveyor William Urquhart. It was not at Golden Point, where they had been working, but on the plateau, with the Yarrowee River creating a division between the town dwellers and the diggers. As the eastern side of the river had not been surveyed, no land could be sold. Nor could businesses be established, though this regulation was ignored as stores and workshops flourished among the diggings to the annoyance of the businesspeople in the town who had had to buy the land on which their businesses stood. Initially there was a lack of water, and then too much of it flooded Golden Point, driving the diggers to higher ground at the base of a spur of the great Dividing Range called Brown Hill. Here on this new development on the Ballarat goldfields, which would become known the world over as Eureka, they discovered that the layers of mud, clay and gravel they found at the bottom of their shafts hid a river which had been formed by water cutting its way through the rocky underbelly of the land many thousands of years ago. Wherever the river changed course or met an obstacle it deposited some of the minerals it had gathered along the way. The volcanic activity that had created Mount Buninyong to the south, Warrenheip to the north, and the plateau on which the town sat had caused it to dam and spread out. In time, all its tributaries had become covered by silt, mud, clay and other debris. New water courses containing fragments of gold developed on top of this accumulation. The Wathaurang, an indigenous clan who came to occupy the region some time in the last 40,000 years, took no interest in the shining flecks they must have seen. Nor did the shepherds employed to tend the sheep of the young Scottish entrepreneurs who, in 1837, had taken up the region under generous pastoral leases, which cost them £10 a year. But the diggers, who had already extracted the gold caught up in the pipe clay layer, knew that there was much more to be found if they could follow the ancient riverbed. It was not long before other buried rivers, or deep leads, were discovered along the stretch of land, which lay below the basalt plateau.

Finding these deep leads presented many problems. It could not be assumed that they had followed the same course as the current streams or even the pipe clay layer. It was a case of trial and error, which could not be done by one man working on his own. By pooling their resources and their claims, four or five men could work together sharing the labour and the rewards. Each person still had to have a current licence, even though it sometimes took up to six months to sink a productive shaft. Some men could support themselves on the gold they had already found, but for others the going was tough, especially as they were constantly being made to produce their licences for inspection by the goldfields police.

Late in 1852, there were only about 2000 diggers at Ballarat, the hard work of sinking shafts keeping all but the most industrious at the numerous shallow goldfields where gold could still be found with a panning dish and a shovel. It was not until the beginning of the following year that the flood of immigrants began to make their presence felt and the population of Ballarat grew rapidly. As they had paid their own passage and expected to return home once they had made their fortunes, they did not consider themselves immigrants. The majority were from Britain, many of them middle class or better, as the poor could not afford the passage price. Men from Germany, France, Italy and Switzerland as well as from North America joined them. The Chinese were not numerous on the goldfields until after the Eureka rebellion was over.

A multinational community

To a remarkable extent, these diggers lived and worked together in national groups. People who came from the same city in England, Ireland or Europe, or were of the same nationality, would be found working side by side, and would often indicate their presence as a group by flying a flag to denote their origin. Although they were not the first or the only ones to stake claims there, the Eureka Lead took on a distinctly Irish appearance. So much so Father Matthew Downing arrived in Ballarat in November 1852 to become the first resident priest on the goldfields, establishing the first Catholic Chapel in a tent at Brown Hill. He later moved it to higher ground on the Melbourne Road, but it was still close enough to his Irish congregation.

As all involved in deep lead mining resigned themselves to remaining in Ballarat for some time, they began to erect more substantial dwellings; many brought their families to live with them. They petitioned for land on which they could grow produce and raise some livestock but this, they were told, was not possible as no subdivision of land could take place until surveys had been completed. In the meantime, the diggers had to make do with a diet of mutton, damper, pickles and black tea all purchased at exorbitant prices from the shopkeepers, who had, in turn, paid dearly for the supplies they sold in their shops. The failure of the government to send surveyors out so that the diggers could legally occupy the land on which their dwellings stood became another one of their grievances.

They also could not get a drink because the government was of the opinion that the sale of alcohol would increase the incidence of crime and disorder on the goldfields. But the drinking of ale and spirits was a tradition, particularly among the working class, in the countries from which most of the diggers came; they resented the prohibition. The more enterprising of the shopkeepers circumvented this by selling sly grog, often from thinly disguised lemonade tents.

In some places, the goldfields commissioners and their police were prepared to turn a blind eye to the sale of alcohol provided sufficient money changed hands in the right direction. In other areas, the authorities smashed kegs and arrested the operators and drinkers alike. By early 1854, the governor had to concede that prohibition had created more ills than it had prevented, and it was replaced by a system of licences for the legal dispensing of alcohol. As the goldfields administration was given the responsibility of issuing these new licences, the corruption which had accompanied sly grog selling was simply transferred to the hoteliers. Those who were prepared to offer considerations to the commissioners and the magistrates on the goldfields were the most likely to receive licences, and be given favoured treatment by the police. They could use the police to put their opposition out of business. Because of their relationship with the government camp, many of the new hoteliers were unpopular with the diggers who continued to frequent the sly grog sellers.

One such hotelier was James Bentley, owner of the Eureka Hotel, whose actions together with those of the police magistrate became the catalyst that galvanised the diggers in Ballarat during the last months of 1854. They were no longer prepared to put up with a corrupt administration, and they wanted other grievances addressed. Most of all, they wanted the rights and privileges which were enjoyed by other citizens of the colony. They wanted the right to vote and to select their own members of parliament who would represent their interests. Nothing less would do. They demanded democracy and were prepared to rebel against the authority of the goldfields and the Crown to gain it. Their rebellion culminated in the brief battle at the Eureka stockade.

Chapter 2

RUMBLINGS

The right of all men to have a say in how they were governed was a concept which had yet to gain acceptance anywhere in the world in 1854, except in the United States of America where it applied only to men of European origin. Elsewhere the right to vote was a privilege handed to men who had demonstrated by their acquisition of wealth that they had reached the necessary refinement to exercise it effectively. Poor men with no education were not entitled to choose who would represent them in the parliament as they were not in control of their own lives. It was believed that they needed the superior intelligence of the upper classes to make decisions for them.

An unequal franchise

The growth of a strong middle class in Britain had necessitated a rethinking of the restrictions on voting rights during the first half of the nineteenth century. By the time Victoria was becoming a separate colony of the

Empire, the franchise had been extended to tradesmen and small business owners. Nevertheless only 10,000 out of the colony's adult male population of 29,000 could participate in the election of the first Legislative Council. Hardly any of them could have put themselves forward for election to the parliament because that office was restricted to men who, by the size of their land holdings or their bank balances, were considered capable of making laws which affected the entire population. As they would not be paid for the privilege of sitting in parliament, they needed to be wealthy to support themselves in a manner befitting their position.

The diggers were certainly not considered worthy of the franchise because of their itinerant lifestyles and the precarious nature of their income. By the middle of 1852, however, many had been resident in Ballarat for up to six months, and were likely to remain there as long as the deep leads continued to yield up their gold. And as gold mining became the legitimate occupation of a large proportion of the colonial population, the diggers began to argue that they had just as much right as the shopkeepers in Melbourne to the vote. Their arguments fell on deaf ears. Besides, the vote alone would not be enough to satisfy them. They also wanted representatives who knew and understood the conditions under which they lived and worked. But the electoral boundaries had been drawn up to suit the pastoralists who, having been the first European occupiers of the Port Phillip District, now the colony of Victoria, considered their interests to be of paramount importance.

Melbourne, with one-third of the colony's population, got only three of the twenty seats that would make up the Legislative Council. Geelong, which was a thriving port and gateway to the western interior of the colony, got two. In both places there were some thousand voters for each politician. In the electorate of Lodden, however, there were only 56 voters for each of the two politicians yet it took in most of the goldfields to the north-west of Melbourne and was teeming with diggers. The other electorates were much the same, a situation that would remain until electoral reform, precipitated by the events of Eureka, brought manhood suffrage to Victoria.

Political impasse

With few exceptions, the men elected to the first parliament showed little flair for government. They were there to protect their own commercial interests and to limit the harm done to them by the discovery of gold. But in this they were thwarted by the Lieutenant Governor Charles La Trobe and his Executive Council, a small group of the most powerful men in the colony, handpicked rather than elected to their exalted positions. According to them, all treasure troves belonged to the Crown, and the lieutenant governor, as a representative of the Crown in Victoria, must out of necessity take charge of all gold-related revenue until informed otherwise by the Queen and her parliament in England. This included the money raised by the sale of licences.

La Trobe would not relinquish any of this money to the Legislative Council for the ordinary running of the colony, insisting instead that it be used entirely for expenses related to goldfields administration. While they waited for advice from the Colonial Office in London, the colony slipped into grave financial difficulty as all other commercial activity on which government revenue could be raised went into decline. In retaliation, the Legislative Council voted to deny the lieutenant governor access to the meagre funds it could raise from the ordinary business of the colony. The stalemate lasted until March 1852, when advice was received from London that all income of the colony was to be included in general revenue. Until then wages of the public service could not be increased to a level which would discourage workers from leaving to try their luck panning for gold. Nor was it possible to employ the additional police needed to maintain law and order, or to institute adequate supervision of those already carrying out licence hunts on the goldfields. Gross abuses of authority, corruption and intimidation were commonplace as a result.

After March 1852, wages of public service employees earning less than £250 were doubled, but they would have to be increased again before the year was out. At the same time, the scarcity of foodstuffs and other commodities meant that prices continued to rise, making recruitment difficult even at inflated wages with only the lazy, incompetent and the corrupt willing to accept positions in the police service either in the towns or on the goldfields.

Protecting themselves

It is not surprising that the diggers had little faith in the government or the goldfields administration. They could not rely on the police to protect them or their property from thieves and rogues who were beginning to make their presence felt on the goldfields. Instead, they felt the need to organise themselves for their own protection. Their meetings were an opportunity for the most outspoken among them to put forward radical ideas like voting rights for all, payment of politicians so that it would be possible for one of their own to represent them in the parliament, and the replacement of the licence with an excise duty which would tax the product of their labour rather than the labourers. Some organisations like the Ballarat Gold Diggers' Association sought to draw attention to the mood of the diggers by sending petitions to the lieutenant governor listing their grievances. They even offered suggestions as to how the administration could be improved, but they were invariably ignored and the petition writers were labelled as rabble rousers.

The diggers were incensed at being referred to as rabble. While there were undoubtedly occasions when sections of the goldfields population were rowdy, particularly when the grog was flowing, and brawls over disputed claims were frequent, on the whole the goldfields were remarkably peaceful. None more so than Ballarat where theatres were beginning to spring up among the tents to provide the diggers with entertainment and a lending library flourished, a sure indication of the level of literacy and sophistication among the digging community.

Fear of Foreigners

This sophistication worried the government. Many of the most literate were foreigners whom the authorities feared would indoctrinate the less well-educated Australian and English diggers with notions of democracy, republicanism and the overthrow of British colonial authority. The number of Irish on the goldfields alarmed the politicians.

Ireland was Britain's most troublesome colony. Seven hundred years of occupation had not broken down the deep hatred and resentment which existed between the Irish and their colonisers. Insurrection was never far from the surface and actual confrontation was frequent. The leaders of the latest uprising had been exiled to Van Diemen's Land, although by the middle of 1853, some of the Young Irelanders, as they had styled themselves, had already escaped with the help of their American supporters, and the remainder would soon receive pardons.

Over the years, the convict settlements of New South Wales, Van Diemen's Land and Norfolk Island had received a great many Irishmen who had been convicted of crimes which could be considered politically motivated. Although the incidences of Irish-led convict revolt were rare, the colonial officials were constantly on alert and were known to take repressive action as a precautionary measure. It is not surprising that the goldfields officials viewed with alarm enclaves of Irish miners establishing themselves on the Eureka Lead.

The Irish and Americans

The gold commissioners were inclined to exaggerate the number of Irish on the goldfields and to attribute most of the violence and drunkenness to them. At no time during the early years, though, did they represent more than about one-sixth of the goldfields population, and those who did come had money enough to pay for their passages to Australia. The majority of those fleeing Ireland in the wake of the Great Potato Famine were so poor they could only afford passages to Britain or to America. In 1852, only 6000 of the 200,000 who left Ireland came to Australia. More would come after the Eureka uprising when the Victorian Government instituted a system of assisted immigration from the United Kingdom and Ireland.

Many of the Irish arriving in the early years had professional qualifications, like young Peter Lalor who had trained as a civil engineer in Ireland before migrating in 1852 with his brother Richard. He initially took a job on the railway being built between Melbourne and Geelong, but then joined his brother in business in Melbourne before trying his hand at mining. Although he took no active part in the growing unrest on the goldfields in 1853, he had come from a family used to taking a stand against oppression of ordinary people by governments who misused the power vested in them. His father was one of the first Catholics elected to the House of Commons following the passing of the *Emancipation Act* in 1828, which lifted all remaining restriction from Catholics in the United Kingdom. He resigned to agitate for a repeal of the Union of Great Britain and Ireland. While his methods were non-violent, those of his third son James Fintan Lalor were anything but. He was one of the arrested leaders of the Young Ireland movement. He died after time spent in Newgate gaol in 1849.

Of greater concern to the colonial government were the Americans. As early as August 1852, the Colonial Office in London had received advice from the British Consul in Philadelphia that a group calling themselves the Order of the Lone Star intended to introduce Australians to the benefits of republicanism. La Trobe was sent instructions to be on the alert for any such activity and to instruct goldfields officials to keep a close watch on all foreigners on the goldfields. Police spies were ordered to attend meetings of diggers and to take down exactly what each of the speakers said. Anyone found advocating sedition was to be arrested.

While the number of foreigners never exceeded more than a tenth of the goldfields population, the Americans did stand out. They involved themselves in all manner of business, not just mining. They arranged for Yankee clipper ships to bring cargoes of American manufactured goods; household items like fancy stoves, mining equipment, boots, clothing and even wagons, which they sold in stores they opened on the goldfields and in the towns. They revolutionised transport with their coach services and built hotels complete with concert halls to which they enticed entertainers of international repute. On 4 July they celebrated with fireworks and bands playing American tunes, much to the annoyance of the lieutenant governor who disapproved of such displays of American triumph over the British Crown.

Americans took to the platforms of the diggers' meetings, particularly in Bendigo and Castlemaine where tensions were on the boil during 1853. At the end of the previous year, the government had muted the possibility of introducing an excise tax, not to replace the licence, but to be charged on top of it. Then the goldfields police committed an atrocity by burning a tent, which housed several women and their children, in the belief that it was a sly grog business; tempers flared. There was talk of raising a force to march on Melbourne to attack the government, and the diggers started to arm themselves. In the end, calm voices prevailed and they elected a 'People's Commission', one of whom was an Irish–American called 'Captain' Edward George Browne. A petition was drawn up and signed by thousands, but La Trobe refused to accept it because it contained too many foreign signatures.

Red ribbons

The diggers were not to be put off. Following the example of their leader, the squatter and rights activist John Harrison, who insisted on a peaceful retaliation, they embarked on a campaign of non-violent protest. They agreed to wear red ribbons in their hats to signify that they were refusing to purchase licences. As they would be easily identified by the police, they hoped to overwhelm the courts and the gaols, forcing the government to take action.

The government did institute a Select Committee for the Better Management of the Goldfields in September 1853 at which some of the diggers' complaints were discussed. The proposal to reduce the cost of licences and to supplement them with an excise tax was quashed by the mercantile community of Melbourne, who claimed that such a tax was not in keeping with the policy of laissez faire by which governments refrained from interfering in the manner in which men did their business. They believed it would also lead to an increase in smuggling. The request that diggers who had been resident for six months on the same goldfield be given the vote was rejected on the grounds that it was far too dangerous while the goldfields were under the influence of foreigners, secret societies and subversives. Ballarat was particularly singled out for its strong republican sentiment.

It is true that the digger population in Ballarat were eager readers of the Melbourne newspaper the *Argus*, which was widely distributed on the goldfields, and that its editors were highly critical of the government. La Trobe had even written to the Colonial Office in London complaining that the paper espoused republican views and was highly anti-monarchical. He was worried that these views would be taken up by the foreigners and those who had been part of the Chartist movement in Britain.

Goldfields Management Act

In November 1853, the *Goldfields Management Act* was passed by the Legislative Council. Licences would cost £1 for one month, £2 for three months, £4 for six months and £8 for a year, but they would not be transferable from one field to another, and the means of policing them would not change. Protest meetings were held on the various deep leads around Ballarat in November and December 1853. What upset the diggers most was the clause, which was subsequently abandoned, stating that the gold and property of unlicensed diggers would be confiscated.

It was at these meetings that a young Welshman, John Basson Humffray, would make his appearance. He had come to the goldfields via New Zealand and was steeped with the arguments of the Chartist movement, which had shaken the foundations of British Government during the 1840s with its demands for manhood suffrage and paid politicians. He was an advocate of winning change by negotiation rather than force, and he soon developed a strong following. While the diggers continued to explore the labyrinth of leads beneath the surface of Ballarat and reap the riches they yielded, they had little desire to engage in anything but peaceful protest. Besides, at least 80 per cent of them held current licences.

Chapter 3

MISMANAGEMENT

In the space of a month in July 1851, Victoria had gone from a district of New South Wales where sheep had greatly outnumbered the 5000 people who made up the population to a seething, rushing, exciting outpost of the Empire where fortunes could be made with a panning dish and a shovel. Some hoped the rush would be short-lived and everything would return to normal. The squatters, as the pastoralists were known because of the manner in which they had acquired their land, would once again be in control, and the population would resume its obedient and subservient position. When this didn't happen, dire predictions about the fate of Victoria began to circulate. The Chief Justice of Victoria prayed for deliverance from the 'unholy hunger' that had inflicted the diggers while most of the law-makers thought it proper to penalise the diggers whose greed had driven them to abandon their duty to their masters in their search for gold.

Unfortunately penalties have to be policed to be effective but at the time of separation Victoria had only a handful of police and 44 soldiers, hardly enough to look after the defence of Melbourne, so the native police were called into service as a temporary measure. Replacing them, however, proved almost impossible at first, and was made more difficult because the needs of the goldfields were considered so different from those of the town; Two separate forces were deemed necessary.

Police

In Melbourne and Geelong, the police were to be modelled on the London bobby, a highly visible and unarmed force whose members kept crime under control by their friendly, but authoritative, presence. Lack of funds to purchase distinctive uniforms meant that they were not nearly as effective as they should have been. On the goldfields the force was styled on the Royal Irish Constabulary, a paramilitary force trained to deal with a population that was recalcitrant and frequently seditious. It's not surprising that the diggers, who considered themselves neither, were offended by their presence and their behaviour on the goldfields. One, in particular, who was shocked by the harassment they inflicted on men who had committed no crimes, nor broken any laws, was Raffaello Carboni, who had left his native Italy in 1849 after fighting for the republican cause in the army of Garibaldi. His experiences, his linguistic ability and his abhorrence of the despotic rule of the gold commissioners would put him at the forefront of the events which came to a head on the Eureka Lead at the end of 1854.

As neither force was attractive to men who could be out digging for gold, La Trobe had to send to London for both police and soldiers. In the meantime, he had to make do with some 130 military pensioners on loan from Van Diemen's Land, but they were far from satisfactory. They were habitually drunk, lazy and ill-disciplined and were returned to Hobart as soon as the four companies of the 59th regiment and a volunteer force of London Police arrived at the end of 1852.

The arrival of the London police did not solve the problem of licence enforcement on the goldfields so incentives were needed to attract recruits. On top of their pay, they were given a half share of all fines their arrests generated. This, of course, only encouraged abuses by goldfields police, many of whom were ex-convicts from Van Diemen's Land, who saw an opportunity of getting rich without getting their hands dirty. It was not uncommon for members of this force to leave the goldfields after only six months with over £1000 in their pockets, an enormous sum for the time, and much more than most diggers could hope to earn with a year or more of hard labour.

Some of the excesses in police behaviour were due to the failure of the government to institute clear-cut chains of command. While the commissioners were responsible for goldfields administration and were answerable to the chief commissioner in Melbourne, the police were under the jurisdiction of the police commissioner and, on the goldfields, the police magistrates. Friction between the two could result in police activity going unchecked on one hand or stifled on the other. Sometimes the commissioners and the police magistrates colluded to cover up abuses, and there were occasions when commissioners were forced to also act as magistrates because no one could be found to fill the position. Despite its size and importance, Ballarat was without a magistrate for much of 1852.

The commissioners

Some of the commissioners had been Crown Land commissioners before gold was found. Others were failed squatters and some were retired British army officers who had served in India. Once the rush from overseas started, the lieutenant governor could also call on young men from good families who arrived with letters of recommendation from British politicians, military officers and members of the aristocracy. Some were very young like 22-year-old Joseph Panton who had charge of the Bendigo field during the Red Ribbon uprising. They were provided with uniforms heavily adorned with brass and braid giving them a quasi-military appearance, adding to their sense of importance.

As the government camp at Ballarat was situated on the plateau where the town had been laid out, it overlooked the goldfields. The diggers' homes and the stores, sly-grog shops and other businesses were all scattered about the diggings in clusters, along the original road from Geelong, beside the Yarrowee River, and up the Melbourne Road towards the Eureka Lead. The diggers would climb the hill to the camp only to buy their licences, unless of course they were dragged there by the goldfields police to answer for some infringement.

The first commissioner at Ballarat was Francis Doveton, an autocratic former military officer whose behaviour was much more in keeping with the commandant of a penal settlement. He was aided by David Armstrong who was a treacherous bully. Originally a blacksmith, he had been a Crown Lands bailiff before being sent to Ballarat as assistant commissioner. He went on to be a police superintendent at Bendigo and then Heathcote. For two years he terrorised the goldfields with a mixture of savagery and intimidation until a public outcry forced his dismissal. By that time, he had amassed £15,000 in bribes and extortion, an enormous amount of money for a man whose salary was £700 per year.

As well as the generous salaries, gold commissioners were provided with horses and servants. On larger fields like Ballarat there were also junior officers of various ranks to share some of the load, and many of these had servants. Their duties were not too onerous for them to engage in social activities like hunting with the nearby squatters, and they dined well on the produce the ordinary diggers never saw. John D'Ewes, the police magistrate in Ballarat in 1854, frequently left the responsibility for the prosecution of unlicensed diggers and criminal offenders in the hands of a justice of the peace when he went on hunting trips.

In three years, the government camp at Ballarat grew from a handful of men, most of them Aboriginal, to over 230 police, soldiers, their officers, and superiors. The commissioner sat at the head of an empire overseeing several assistant commissioners all with their own area of responsibility, their own police, and methods of dealing with the diggers. The cost, in 1854, of their maintenance in wages, clothing and food, and fodder for their horses, was in excess of £100,000 when the income from licences was only £105,000. It was much the same on the other goldfields.

Land

One of the major complaints on the goldfields was the lack of land for purchase. By 1854, some had been made available but in lots which were too large and too expensive for most diggers. Besides, they were disposed of at auction instead of being sold for a fixed price per acre. As a result, squatters — who were the only ones with sufficient capital — snapped up most land lots. To add to the distress of the diggers, most of this newly purchased land was left idle while they and their families went without the produce it could have provided.

Ballarat was changing in other ways, though. Timber had replaced canvas as a building material, and by early

1854, when prohibition was lifted, a number of hotels had sprung up. The first to be established was Bath's in Lydiard Street on the edge of the plateau. It would eventually be sold to Walter Craig and become the landmark Craig's Hotel. Another, built conveniently opposite the government camp, was the George. Some of the hotels on the flat also had dining rooms, music halls and bowling alleys, like the Eureka built by an ex-convict called James Bentley on the Melbourne Road just near the Eureka Lead.

By 1854, the population of Victoria had grown to 250,000 of whom about 25,000 were living in Ballarat. Lack of suitable labour for the colony's business and industry had been replaced by massive unemployment as new arrivals flooded in only to find they did not have the funds to get themselves to the goldfields or a liking for the life once they got there. Most had come from the industrial cities of England and had few skills that would gain them employment in Melbourne, although some were absorbed into the goldfields police.

Charles La Trobe

Charles La Trobe had been at the helm for fourteen years, having been appointed superintendent of the Port Phillip District in 1839. In the last three years of his rule he had seen public expenditure increase from £75,000 to £1,225,000 and gold surpass wool as the major export of the colony. But the government he led had also been responsible for soaring inflation, making Melbourne the most expensive city in the world to live in. There was a deficit of £2 million and imports had exceeded exports over the previous two years by a staggering £10 million. The stores were full of commodities and foodstuffs, some of them perishable, which could not be sold because the population could not afford them.

In December 1852, La Trobe tendered his resignation to the Colonial Office but was unable to leave until a suitable replacement could be found. Twelve months later he sent his wife Sophie home so that she could have medical treatment in Switzerland, and began to make preparations for his own departure. Before he could leave, there was much to be done. The *Goldfields Management Act* had to be sent to London for ratification. In it he included a strong recommendation that Her Majesty give immediate assent to a separate bill to enfranchise the most settled of the mining community. Unfortunately her reply did not come until after blood had been shed at Eureka.

La Trobe left Melbourne on 6 May 1854, one day after he read in the newspapers about the death of his wife.

Chapter 4

PROVOCATION

The diggers were delighted to hear on 22 June 1854 that Sir Charles Hotham had arrived to be the next lieutenant governor of Victoria. News of his distinguished naval service had preceded him. He'd fought pirates, broken blockades and negotiated treaties on behalf of the British Government in South America. He was just what the colony needed, a strong, decisive man who could bring the colonial government to order, rein in the excesses of the goldfields administration, and bring justice to all.

What the diggers didn't know was that Hotham was coming to them reluctantly. The governorship of a small troublesome colony at the end of the earth was poor reward for a man with his distinguished record of service to his Queen and the British Empire, especially at a time when Britain was embarking on major conflict with Imperial Russia in the Crimea. Much persuasion had been necessary before he'd accept the position. He had demanded, and been granted, a salary of £10,000, which was five times that of his predecessor, and he'd been assured of an adequate supply of troops should he need to use force to assert his authority over any segment of the colonial population. Five companies of 50 men had been newly despatched to the colony, and he had permission to call on his neighbours in New South Wales and Van Diemen's Land if he needed more.

He was newly married to the grand-niece of Lord Nelson, and had no intention of bringing his bride to the humble prefabricated house that the La Trobes had brought with them to the Port Phillip District in 1839. A mansion was purchased in Toorak but it failed to meet his expectations. It was too small and the furniture it contained was fit for a pigsty. While the floor in the stables was thick with mud, there was an extremely well-stocked wine cellar, the contents of which he sold back to the supplier.

Touring the goldfields

While he had the officials quaking at the knees with his demands and complaints, the people were gaining a different impression of him. In Geelong, their hopes were raised when he told them he believed that all power proceeded from the ordinary people. In Ballarat on 26 August 1854, he walked among the shafts in the Gravel Pits, a recently discovered lead between the original Golden Point and Black Hill, another of the spurs of the Great Dividing Range. He listened to the diggers' complaints, watched them work, and inspected the lumps of clay studded with gold they brought to the surface. He was impressed by their manner, their ingenuity, and their physical skills. He understood the risks they took and the disappointments they faced if their months of labour brought no result. He believed them to be loyal and honest men who deserved to be treated with fairness. Even the normally critical Henry Seekamp, editor of the recently formed *Ballarat Times*, was impressed. A 'bold, vigorous and farseeing man has been amongst us', he wrote. Seekamp was confident that the useless restrictions under which the diggers were shackled would soon be abolished.

The diggers too had treated the lieutenant governor and his good lady with great respect, one even lifting Lady Hotham over a particularly nasty puddle. They had proudly shown them samples of the gold-encrusted rock and clay, which they'd brought from the holes they had dug deep into the ground, and a nugget weighing 35 kilograms found that day had been called the Lady Hotham in their honour. Unfortunately, this generosity was to be their undoing as it only served to confirm in the mind of the lieutenant governor the affordability of the licence. This was the message he delivered when his tour of the goldfields took him to Bendigo. Liberty and order came at a price. The diggers could not expect to enjoy harmony on the goldfields unless they were prepared to pay for it.

Having seen the goldfields in operation, Hotham's immediate task was to deal with the financial mess he had inherited from La Trobe. He had also inherited an

Executive Council he believed he could not trust, a Legislative Council who had little talent for government, and a public service that was clearly incompetent. While he may have wanted to sweep Victoria clean with a new broom, he could not replace an elected government or the Executive Council at short notice. He chose to ignore most of them and run the colony the way he had run the ships once under his command: autocratically. He even kept the attorney-general and the colonial secretary at arm's length, issuing orders that had to be obeyed without discussion. As trouble brewed in Ballarat he bypassed the chief commissioner of the goldfields. He carried on a direct, private, and frequently secretive correspondence with the resident commissioners.

He had good reason to be unimpressed with his public service. Confident that the diggers would eagerly embrace the new one-year licences, they overestimated by a staggering £100,000 the likely revenue for the first six months of 1854. Unfortunately, the longer licences provided no incentive for the diggers. The one-year licences were four times the price of the three-month licences and, being non-transferable between goldfields, would cost them dearly if they decided to try their luck on another field. Since the Red Ribbon unrest of the previous year there had been a steady decline in the numbers taking out licences despite the cost reductions. At the beginning of 1854, only 65 per cent of diggers were estimated to have bought them, and by the middle of the year compliance was down to 50 per cent. This, Hotham believed, was due not to the diggers' intransigence, but to the goldfields officials who had grown fat and lazy in the execution of their duty. It was their responsibility to impress on everyone under their authority the necessity of having a licence, and to punish all who disobeyed.

Licence hunts

On 13 September 1854, ignoring all advice given to him that a further souring of relations between diggers and the police could have explosive consequences, Hotham ordered that licence hunts were to be increased from two or three times a month to twice weekly. The police, of course, relished the opportunity to dramatically increase the money they made on fines and took to the task with great enthusiasm. All around the goldfields the warning cries of the diggers could be heard. 'Joe! Joe!' they shouted whenever the sound of horses' hooves signalled the approach of the licence hunters, the call having been adopted in the early days because it was the middle name of Lieutenant La Trobe.

On Ballarat goldfields, where the shafts were the deepest, the diggers were battling the ever constant problem of underground water, and the necessity of climbing the 31 to 46 metres to the surface caused them great difficulty. If they had to be away from their holes for any time at all and were unable to continue bailing the water out, not only would their claim flood but also all those next to it would be in danger. This was of no concern to the police or the assistant commissioners who each had control over a lead or district of the goldfields. And as the boundaries of these districts were ill-defined, it was not uncommon for police from one to cross over into the territory of another, thereby forcing diggers out of their holes twice in one day. Anyone found without his current licence on him was arrested, even if he could prove that he had taken one out.

Police unpopularity was dealt a further blow when a digger, arrested for drunkenness, bled to death while in custody. It appeared he had cut himself on one of the thousands of bottles which littered the goldfields; he had been refused treatment until it was too late. Then an American restaurateur had been imprisoned on a trumped-up charge of dispensing sly grog because he would not pay a bribe to Sergeant Major Robert Milne, who had been profiting well from an extortion racket he was running in Ballarat, where sly grogging continued to flourish despite the availability of legally dispensed alcohol in the hotels. Those who paid Milne were allowed to continue to operate while those who didn't were arrested, their tents burned to the ground, and their casks taken for police use or staved in. The American in question did not sell liquor. His case was taken straight to the lieutenant governor by the American consul in Melbourne and he was released, but Milne was not brought to justice until he had further infuriated the diggers by giving evidence in defence of James Bentley at the inquest into the death of James Scobie.

First death on Eureka

Bentley had come a long way since being transported to Van Diemen's Land in 1843 for forgery. He'd worked as a confectioner and gold buyer in Melbourne before opening a general store in Ballarat in 1853. By the following February, he had acquired enough friends in high places to support his application for a liquor licence. He then set about building his hotel called the Eureka, on the Melbourne Road. It was a most palatial affair with chandeliers, three bars, gaming rooms and sleeping accommodation for 80 people; there was also a bowling alley at the side of the hotel run by an American. It was rumoured to have cost £30,000, much of which Bentley had borrowed. His benefactors included the police magistrate at Ballarat, John D'Ewes, who was widely believed to have been a part owner, and a man called Farrell, an ex-convict like Bentley, who had taken advantage of the recruitment difficulties of the previous administration to make himself very rich as police magistrate in Castlemaine. He had amassed a fortune of £15,000 in fines and bribes before he had conveniently disappeared.

In the early hours of 7 October 1854, Farrell was dining with James Bentley and his wife Catherine when a young Scot called James Scobie knocked on the door of the hotel looking for a drink. He'd been out all night celebrating a reunion with his friend Peter Martin, whom he had not seen since leaving Scotland. As they were both quite drunk they were turned away. Words were shouted, and Scobie retaliated with some insulting remarks about Mrs Bentley. A stone was thrown at a light. At Martin's urging, they left and began walking up the road towards the Eureka Lead where Scobie had his tent. They heard voices and turned back on the chance that Bentley's barman had had a change of heart. They were met in the dark by four men and a woman. At least one of the group had a spade. Martin was struck to the ground but was able to get to his feet. He ran back to his friend left lying on the road. He had been attacked with the spade and kicked as well. He was dead by the time he was carried inside the hotel by the doctor who had his practice nearby. The death of this young Scottish digger triggered a chain of events that culminated in rebellion at Eureka.

Chapter 5

CORRUPTION

By the time the sun rose on the morning of 7 October 1854, news of Scobie's death was all over the goldfields, and those who were able made their way up the hill to the government camp that afternoon to be present at the inquest. A local doctor, David Williams, acted as coroner. He began by taking evidence from James Bentley who claimed he had not left his hotel all night. He'd no knowledge of the young Scot's death until told of it by the doctor who had brought the body in off the road. He didn't mention that the doctor, Alfred Carr, was a friend of his and a regular drinker at the Eureka. The barman told a similar story, as did other members of Bentley's staff, but a very different one was told by a little boy called Bernard Welch.

Bernard lived with his family in a tent almost opposite the hotel. As the night had been very warm the flap was open, giving him a clear view of the road outside. Woken by shouting, he looked out to see Mr and Mrs Bentley and three or four other men standing on the corner by the tent. One of the men picked up a spade before they all walked on briefly. There was a scuffle and he heard a blow being landed.

It was then that the coroner showed his true colours. Despite protests from some of the jurors he allowed Bentley to cross-examine the boy, who stuck to the evidence he gave. As proceedings continued, with Dr Carr giving his opinion as to the cause of death, Williams must have been aware that Bentley was sitting among the jurors whispering to one of them. Further conversations took place between them after the jury had retired to consider the verdict. It was no surprise to the diggers that death by persons unknown was the outcome. They would not let the matter rest, however. They demanded that Bentley, his wife and associates be charged with the murder of Scobie and that a proper inquiry be held. They got their wish. Two days later, Bentley, his wife, Farrell and the barman were arrested and bailed to appear before a judicial inquiry set down for 12 October.

Poor justice

This brought the diggers no satisfaction either because one of the three members of the board of inquiry was none other than Police Magistrate D'Ewes, who clearly had a vested interest in seeing Bentley acquitted. He was joined on the board by Resident Commissioner Robert Rede and Assistant Commissioner Johnston. Rede came from a distinguished naval family and had studied medicine before coming to Victoria. After dabbling in mining for a while he was appointed a commissioner, and then resident commissioner of Ballarat. Initially his sympathies were with the diggers, believing that the licence system was the root of all evil on the goldfields, but his enforcement of it nevertheless lost their respect. They could not trust him.

Johnston had also recently incurred the wrath of the diggers for his handling of the wrongful arrest of the Catholic priest's servant, a crippled Armenian who had been ordered out of the tent of a sick man he had been visiting and dragged off to the lock-up. Instead of explaining to the constable that clergy and their servants did not require licences, Johnston had him charged with assault and fined £5, which the priest had to pay. His actions infuriated the Irish who met after Mass on the following Sunday to draw up a petition for the removal of the constable and Johnston from the goldfield. Like so many other petitions which were sent to the lieutenant governor over the years, it fell on deaf ears. Hotham did not believe that the assistant commissioner had acted improperly, but to deny the Irish a reason to feel aggrieved, he did suggest Johnston be moved to another goldfield. Rede refused as he had no one to replace him. However, it was fortunate for the diggers that he was able to take his place on the board because he was the one dissenting voice to the decision the other two members reached that Bentley had no case to answer.

Right from the outset it was clear that D'Ewes was determined to have the charges against the Bentleys dropped. He interrupted the lawyer for the prosecution and he cross-examined the witnesses, particularly little Bernard Welch who did not waver in telling what he had seen. By now, there were more witnesses who had either seen or heard Bentley and his wife in the street at the time Scobie was killed. A doctor called by Scobie's brother had even matched the spade with the blows on the dead man's skull. The lawyer for the defence was able to produce just as many witnesses who could swear that Bentley never left the hotel on the night in question. During an adjournment, Bentley was seen to be deep in conversation with D'Ewes in the police magistrate's room. Shortly afterwards the verdict was announced, but Johnston did not agree with it or the manner in which the inquiry had been conducted and he said as much in a deposition he sent to the attorney-general in Melbourne.

The burning of Bentley's Hotel

Angered by Bentley's acquittal the diggers decided to meet on the Tuesday after the inquiry at the place where Scobie met his death. It began in an orderly fashion with a committee being elected to press the case for a retrial of the alleged murderers. As there had been an order from the chief commissioner that all such meetings were to be monitored by the police magistrate and two witnesses for evidence of sedition or republican sentiment, there was a contingent of police present, but D'Ewes was not there. In his place was an inspector called Evans. As the heat of the day increased, the diggers began moving down the road towards the hotel. They demanded Bentley show himself but he had escaped through the back entrance and was making his way at full gallop on a trooper's horse towards the government camp. Tempers were raised and there was much shouting. Some small boys in the crowd began throwing stones; one hit a light. A window was smashed, and another, and some of the diggers climbed inside. They began throwing out furniture, bottles and casks. Some of the liquor was drunk on the spot, the rest poured out as they were not there to loot but to destroy everything the murderer of Scobie had owned.

When word of the disturbance reached the government camp, Rede set off with the intention of reading the *Riot Act*. He managed to climb on to a windowsill as the weatherboards were being torn from the building by the angry diggers but before he could make himself heard he was pelted with eggs someone had stolen from the hotel kitchen.

Then a cry of 'Fire' went up. The fire started in the bowling alley, rapidly spreading to the hotel. The soldiers refused to attempt to extinguish it as fire-fighting was not their duty. They retreated while the police made a valiant but futile effort to quell the flames. They too returned to the government camp leaving the hotel to burn to the ground. They found Bentley and began to panic, knowing that they could not defend the camp against an onslaught of thousands of diggers. The women and children were immediately removed to safer quarters and 1000 rounds of ball cartridges were issued. Soldiers were put to work increasing the fortifications but they need not have worried as no attack came.

The following morning, a letter written by John Basson Humffray appeared in the *Ballarat Times*. While he deplored violence of any kind, he believed that the blame for the burning of the hotel rested squarely with the government officials, who had goaded the diggers into action by failing to administer the law properly. Unless they were prepared to address the abuses under which the diggers suffered, he feared there would be more violence. The committee, elected the day before, sent off a petition again asking that Scobie's killer be brought to justice.

More arrests

Hotham had already posted a reward for the killers and, acting on Johnston's deposition, had instigated a board of inquiry to look into the manner in which the law had been administered in the Scobie case. In the meantime, Rede had given orders that two men were to be arrested in relation to the riot and the burning of the hotel. As thousands had taken part in it, the unlucky two had to be chosen at random. One had certainly been there but whether he was encouraging his fellow diggers or trying to restrain them no one was sure. The other man had been nowhere near the hotel. D'Ewes set them bail of £100, which was paid with money collected by the diggers' committee. Rede intervened and demanded it be increased to £500 with a £250 surety for each of the men.

In response to Rede's contention that the possibility of confrontation between the government camp and the diggers was growing more likely with every day, reinforcements were sent. A detachment of mounted and foot police had arrived on Thursday, 19 October. They were followed by more men of the 40th regiment and another detachment of the 12th regiment who joined the detachment already based in Ballarat. In addition, D'Ewes attempted to swear in special constables but there were only three applications and none of them were diggers.

The Diggers' Reform Society

The diggers were meeting at Bakery Hill, a small rise between the Gravel Pits and the original Golden Point. They realised that they needed to be more organised and orderly in their opposition to the goldfields administration and the government in Melbourne. They could not allow themselves to be branded a riotous mob, or give Rede any excuse to read the *Riot Act* that would allow him to send in troops to suppress them. They would use Hotham's own words to press home their message — that the power came from the people and that they had been provoked by the actions of the officials into the violence which had taken place — and they would demand the exoneration of the men who had been arrested. They all agreed to the formation of the Diggers' Reform Society giving Humffray, his fellow Chartists Kennedy and Holyoake and a handful of others the authority to put together a charter they would all support.

Rede knew what they were doing. He had spies in the crowd who reported on every word, and added a few of their own as it suited them to exaggerate the danger the government camp was facing. He decided that the Diggers' Reform Society represented an escalation of the state of insurrection existing in Ballarat, and that it had to be crushed at all costs. In the meantime, he had his police arrest more men in relation to the riot and the burning of the hotel, although only one of them was brought to trial.

Chapter 6

MOUNTING TENSION

As well as offering a reward for the capture of Scobie's killers, Hotham ordered the arrest of Bentley, his wife and his associates. This prompted some of the past employees of the Eureka Hotel to reconsider the stories they had given previously. It appeared that it was Farrell who had picked up the spade outside the Welches' tent that night, and it was he who had struck Scobie the fatal blow. He was also the one who kicked the poor Scot about the head and body once he was down. The barman and Mr and Mrs Bentley had looked on, and then they had all returned to the hotel leaving the injured man where he lay.

The Bentleys had fled to Melbourne and took some finding. Eventually, though, they were apprehended and committed for trial on 18 November 1854 at the Supreme Court in Melbourne with the formidable Justice Redmond Barry presiding. This meant that all the witnesses had to be brought down from Ballarat, which proved quite difficult, as some of them had moved on.

Fortunately, they were found in time to give evidence in the court. Young Bernard Welch was accompanied by his mother, who gave evidence about Mrs Bentley's role in the crime. She had heard Mrs Bentley say to the men she was with, 'That is the way to treat these sweeps', just after the fatal blow landed on Scobie.

Bentley and his co-accused were well represented by some of the best lawyers in the colony, who were so persuasive that the jury, on the advice of the attorney-general, agreed to the lesser charge of manslaughter. Bentley, Farrell and Hance the barman were all sentenced to three years' hard labour. Mrs Bentley, who was heavily pregnant, was acquitted.

The trial of the rioters

The day after the Bentley trial, the three men who had been arrested for taking part in the riot and the burning of the hotel also faced Justice Barry. The Reform League had used the funds it had collected to employ an eloquent Irish lawyer, who argued that the real villains were the corrupt officials who had acquitted Bentley in October. If they had done their duty there would not have been a riot, and the hotel would not have burned down. He wanted to question Rede but the attorney-general, who was prosecuting the case, would not allow it. Nor would he allow the jury to take into consideration the provocation by the soldiers and the police. The three men were given an opportunity to explain what they had been doing at the time of the riot so the jury could decide the extent to which each was implicated in it. While finding them all guilty, the jurors added a rider, which the attorney-general refused to accept, expressing their belief that they would never have had to do such painful duty if the authorities at Ballarat had done theirs properly. Fortunately one of the jurors kept a copy of the rider, handing it over to a reporter from *The Age* newspaper where it appeared the next day. The newspaper also reported that one of the men was given three months in Melbourne Gaol. Another four men, the last to be arrested, were given six months.

In the meantime, the three-man board of inquiry set up by Hotham to investigate the causes of the disturbances began taking evidence from officials, miners and shopkeepers in Ballarat. They were limited in their questioning to the immediate events and were therefore not able to delve into the real causes of unrest. They did note in the report they issued on the day after the trials finished that the goldfields administration was poor, the police and the military were negligent in their duty to protect the property of businesspeople, and they failed to act quickly enough to prevent Bentley's hotel from burning down. When forced into action they had behaved in a cowardly manner. The police magistrate was underhand in his dealings with licensed victuallers from whom he borrowed money and he had no intention of repaying them as a means of securing their loyalty. He had also perverted the course of justice by acquitting Bentley and his associates.

The Ballarat Reform League

The Diggers Reform Society had also been busy developing a charter, which they presented to the diggers who assembled in great numbers at Bakery Hill on 11 November. Over the previous few weeks, they had rejected all calls to take the law into their own hands. Paramount to their charter was the use of moral persuasion and the resistance of physical force until all avenues for addressing their grievances were exhausted. Heading the charter was a declaration of the right of all citizens to a voice in the making of the laws which affected them. This meant not only manhood suffrage but also the right to select their own representatives in the parliament. This, they believed, should be achieved by the removal of all property qualifications on parliamentary representation. More importantly, there could be no taxation without representation. This statement would become a catchcry of the diggers in the days and weeks to come. The charter was accepted enthusiastically by the diggers, who elected Humffray as the first president of the newly named Ballarat Reform League. George Black, who like Humffray was a Chartist, was elected secretary. All were prepared to acknowledge their loyalty to the colony and to the Crown, but they warned that if the Queen continued to accept the advice of corrupt and dishonest ministers who denied them their rights as her subjects, they would have no alternative but to separate themselves from her government. This was the first serious suggestion of republican intentions.

Government scheming

Hotham, meanwhile, was scheming with Rede as to the best means he could employ to crush the power of the Ballarat Reform League, which they called a secret society although there was nothing secret about it. Its formation and charter had been widely reported in the local and Melbourne newspapers. Even before the burning of the hotel, Rede had taken to employing plain-clothes detectives to spy on the activities of the diggers. Some of the information they relayed back was nothing short of scurrilous. They told of large numbers of diggers undertaking military-style training, and plots to storm the government camp to rescue prisoners incarcerated in the lock-up, all of which were figments of the spies' imagination.

Ever distrustful of the colonial government, Hotham devised a code by which he and Rede would communicate. No one else had access to it, not even the chief commissioner who was completely sidelined from the events which were unfolding in Ballarat throughout November, or the colonial secretary who, though he had considerable sympathy for the diggers and their plight, was made to shoulder much of the blame for the worsening tension on the goldfields. He was forced to resign in the wake of the uprising.

Rede also didn't trust all those reporting to him and began to suspect that just as he was spying on the activities of the diggers, they were being supplied with information about the whereabouts of prisoners, troop deployments, and the fortifications which had been erected in the camp. The resident commissioner was beginning to see himself as a general in charge of a military campaign. There was an atmosphere of heightened anticipation with troops in a constant state of readiness for an expected attack. While he waited for an opportune moment to take the diggers on with his military might, Rede went out of his way to provoke them by stepping up the licence hunts.

For the time being the diggers refused to be provoked. Those with licences produced them when harassed by the police to do so, and they protected those without as best they could. They had faith in Humffray and the Ballarat Reform League that justice would be done and their fellow diggers facing the court in Melbourne would be freed. When they were not, the executive of the league called an urgent meeting to decide their next move. Humffray was for drawing up yet another petition, but others argued that the time for petitioning was over. The diggers would no longer plead or beg, they would demand their rights and they would deliver their demands in person to Hotham.

Demands

The meeting between Hotham and Humffray, Black and another Chartist member of the Ballarat Reform League, Tom Kennedy, was a disaster. Hotham's naval upbringing came to the fore. No subordinate of his in the navy had ever demanded anything of him, and as the lieutenant governor he would not tolerate such insolence in the colony of Victoria. He would not accept the document they had prepared for him, nor would he listen to their explanations as to why they had used such strong language even though he was just as guilty as his predecessor of having petitions filed away unanswered. Even the attempts by the three delegates to modify the tone and remove the word 'demand' fell on deaf ears. They returned to Ballarat fearing they would not be able to prevent bloodshed on the goldfields for much longer.

Others feared the same. The day before the fateful meeting between the lieutenant governor and the delegates, the Catholic priest, a young Irishman called Patrick Smyth, had got wind of a rumour that the diggers were planning an attack on the government camp and that, in the fashion of Irish uprising over the centuries, they were manufacturing pikes out of strips of metal and wooden poles. He had also heard that they had 1000 rifles. He took his concerns to Rede who had had similar information from one of his spies. The editor of the *Ballarat Times* threw more fuel on the fire by announcing in his newspaper that something momentous would happen on the following Wednesday. Rede immediately applied to the attorney-general for permission to arrest him for sedition. He sent his spies out again with instructions to find information about the weapons cache. Despite their best efforts, though, and a considerable amount of money which was meant for bribes but which they probably pocketed, they found nothing. On the Monday morning, the diggers were all hard at work on their claims, and peace reigned on the goldfields. It was far too peaceful for Rede who could barely contain his lust for action.

The momentous event scheduled for the following Wednesday turned out to be nothing more than another of the monster meetings which the Ballarat Reform League planned to have at Bakery Hill so the diggers could be told the outcome of the delegates' meeting with Hotham. On the Monday evening, however, a brief skirmish took place which, together with the bad news the delegates carried, would change the direction that meeting would take, and put the diggers on an irreversible path to confrontation with the government.

First blood

A small detachment of the 12th regiment, which was escorting some wagon loads of ammunition arriving late to Ballarat, veered off the Melbourne Road to stumble their way across the Eureka Lead towards the government camp. They were set upon by a party of diggers from the Gravel Pits who had been angered that a detachment of officers and 106 men of the 40th regiment from Geelong had entered Ballarat a few days earlier with their bayonets at the ready. Shots were fired, a drummer boy was injured and the wagons were overturned, but not before the officers accompanying them were able to signal to the camp that they were under attack. Aid quickly arrived and all the men were able to reach safety, including the drummer boy who was later reported to have died of his wounds. Although a gravestone was made for him he survived to live for many years afterwards.

Despite the fact that the wagons carrying the ammunition were owned by an elderly American who was so aggrieved by their loss that he demanded compensation, Rede now believed the Americans were playing a double game on the goldfields. Even though they did not appear to be playing an active role in the Ballarat Reform League, he had detected a strong republican undertone in the charter which had been produced. He decided he would have his spies attend the meeting, which had been hurriedly called for the afternoon of Tuesday, 29 November.

His confidence in the camp's ability to crush the League had been strengthened by the arrival of additional reinforcements. The soldiers who had come from Geelong were accompanied by a detachment of mounted police. More troops had come up to Ballarat from Melbourne and from Castlemaine through Creswick where they found everything quiet and peaceful. By the end of November, there were 435 police and soldiers within the camp under the command of Captain Thomas of the 40th Regiment.

The flag unfurled

Father Smyth was much less confident about the fate of the diggers, and not knowing what he could do to protect his flock, he wrote to his superior in Melbourne, Bishop James Alipius Goold, asking for his advice. Goold travelled all night by way of Geelong, where he collected Father Matthew Downing who was well known for his ability to calm troubled situations. They arrived in Ballarat at 10am and tried in vain to dissuade the Catholic diggers from going to the meeting. In the end, Goold had to concede that the time for conciliation had passed. He prayed that the meeting would be peaceful but he did not attend it, although the other two priests did take their place on the platform. Above them flew the flag, white stars and a cross on a blue background, made over the previous few days by the women in the district out of material at hand.

Posters printed by the *Ballarat Times* had brought diggers in their thousands from all the local fields to hear what the delegates had to say. It soon became clear that the executive of the Ballarat Reform League were divided as to what should happen next. Some like Humffray wanted to persist with the moral persuasion approach and suggested that they reword the document to remove the offensive word 'demand'. Others believed that it was too late, particularly in light of the attack on the soldiers on the Eureka Lead. Most of the diggers thought likewise. The Chartists, who had held the authority until then, had no alternative but to withdraw.

They would play no further part in the events which were unfolding.

The meeting threatened to break into disarray. There were agitators calling for immediate action, and others were urging the diggers to burn their licences. It was then that Peter Lalor, who until then had been more of an observer than an active participant in the events which had been unfolding, stepped forward and moved that another meeting be held on the following Sunday at the Adelphi Theatre to elect a more representative committee to voice the diggers' wishes. That meeting would never be held; by then the battle had been fought on the barricaded piece of land on the Eureka Lead. The die had been cast.

Chapter 7

STOCKADE

Rede immediately knew about the burnt licences and the harsh words that had been spoken. He also knew that the moderate forces of the Ballarat Reform League had been defeated. This no doubt pleased him as any show of militancy on the part of the diggers would give him the excuse he needed to unleash his much better trained and armed forces. He persisted with the belief that they were being influenced by a handful of agitators who were using the licence issue to promote revolution, and that he only needed to show them he would not hesitate to use whatever means were necessary to get them to capitulate. To test this belief he ordered another licence hunt for the following morning, Wednesday, 30 November. The Gravel Pits were chosen as there was a clear view of this field from the camp should anything untoward happen.

Under the leadership of Assistant Commissioner Johnston, the police set out accompanied by soldiers with their bayonets drawn. They were soon under attack from the diggers who pelted them with stones, sticks, bottles and anything they could get their hands on. The diggers also refused to show their licences and would not disperse when Rede rode among them with great difficulty; his horse, spooked by the noise and flying missiles, kept rearing up. Rede read the *Riot Act* as more troops advanced in skirmish formation across the flat. Shots were fired from both sides, and one policeman and a digger were slightly hurt. Eight diggers were arrested and hauled up to the camp.

When word of the bloody licence hunt reached the other Ballarat fields, the diggers began to make their way to Bakery Hill in the hope that one of their number would know how they should respond to what was certainly an escalation of the violence with which Rede intended to enforce the law. Again it fell to Peter Lalor who, while admitting that he had no military experience, recommended that the diggers take steps to defend themselves. Standing beneath the flag of the Southern Cross, he called on them to form into groups and to elect captains from amongst their ranks. He asked Alfred Black, the brother of the secretary of the Ballarat Reform League, to note down the names of the captains. Lalor turned to his friend Raffaello Carboni, who'd had much experience fighting against much better equipped and organised forces in his native Italy, and asked him to be his right-hand man. His first task was to use his gift for languages to translate Lalor's message to the foreign diggers.

To Eureka

As Bakery Hill could be kept under observation from the government camp, Lalor directed the diggers to follow him to the Eureka Lead, where they would be afforded some protection from the undulating ground. A young Canadian called Henry Ross, who had been responsible for designing and commissioning the flag of the Southern Cross, lifted it up so that the companies of diggers could fall in behind it as they marched towards Eureka.

Once there, Lalor addressed them again. He said nothing about rebellion as his primary concern was that of protection of his fellow diggers from unprovoked attacks by the goldfields administration, and if possible to rescue the men who had been arrested earlier in the day. He drew together the men he could trust into a council of thirteen who would help him formulate a strategy to deal with any future attacks. It included Carboni, his friend Timothy Hayes, John Manning from the *Ballarat Times*, and Patrick Curtin who would be in charge of manufacturing the pikes. The Black brothers were almost certainly involved and there were a couple of Germans, one of whom, Frederick Vern, claimed to have had military training and was most put out when he was not given the title of commander-in-chief, which Lalor took for himself. The rest of the council were English and American; of these, a young man called James McGill was supposed to have been trained at West Point. Lalor was disappointed that the Ballarat

Reform League leaders seemed unwilling to continue, but he pledged himself to the service of the diggers. He told them: 'I tell you gentlemen, if once I pledge my hand to the diggers, I will neither defile it with treachery, nor render it contemptible by cowardice.'

The site Lalor chose to defend was about 200 metres further along the Melbourne Road from the remains of the Eureka Hotel and was about an acre (0.4 of a hectare) in size. Although it was pitted with shallow holes where speculators called shepherds had secured claims they intended to work once they knew the direction the underground streams had taken, it had not been mined much. It already contained several tents belonging to the mainly Irish diggers and their families, and there were a couple of shops. Lalor had erected his own tent there. They were all allowed to remain where they were while a fence made of slabs of timber used to line walls of shafts, tree trunks and whatever material was available was built around the perimeter. Earth was shovelled against this fence to provide extra strength.

One of the men earnestly knocking slabs into place and levelling the ground inside the fence was Henry Goodenough, a police constable who was being paid to spy. He returned to the camp later that day to tell Rede the diggers planned to launch an attack at 4am the next morning. Nothing was further from the diggers' minds. After a hard day's work they returned to Bakery Hill where the Southern Cross was once again unfurled.

Diggers who had not gone with them to the stockade joined them to listen to Lalor, who stood before them, a rifle in his left hand, asking all those who were not prepared to swear an oath to leave. Some drifted away but the 500 who remained fell in behind their captains and joined with their commander-in-chief who went down on one knee, his right hand pointed towards the flag fluttering above, and led the assembled diggers in the first oath sworn in the Colony of Victoria under a flag that was not British. They pledged: 'We swear by the Southern Cross to stand truly by each other, and fight to defend our rights and liberties.' Then they returned to the Eureka.

Over the next two days, diggers were seen coming and going from the stockade. Some were on official business seeking weapons and supplies from the various businesses. They issued receipts for everything they obtained. Unfortunately some of the more unruly elements on the goldfields seized the opportunity to harass storekeepers and loot their supplies. Reports of lawless mobs reached the camp and detachments of police were sent out as Rede wanted a chance to catch the diggers with arms in their hands. Unfortunately for him, none could be found.

December 1854 began quietly enough. It was a Friday and Rede expected to see the men at their claims, but the goldfields were at a standstill. No work was being done. By 11am a crowd had begun to congregate at Bakery Hill and the police were sent to disperse them. The *Riot Act* was read and two more diggers were arrested, one of whom was armed; a sure sign that there was rebellion afoot.

Spreading the word

Unbeknown to Lalor, Kennedy and Black had decided to spread the news that the diggers at Ballarat were under attack by the goldfields administration. They addressed a large gathering of diggers in Creswick, informing them that there were arms and food available to those who wished to help fight for the rights they all wanted. Four hundred diggers volunteered and were led triumphantly into Ballarat singing the French national anthem, the Marseillaise, in recognition of the similarities of their march with the storming of the Bastille, only to find that no provision had been made for them and there were no weapons to spare. A local butcher provided meat to feed the newcomers but most returned to Creswick once they discovered that Lalor still hoped for a peaceful outcome to the dispute, and their services were not welcome.

By Saturday, 2 December, the diggers in the stockade and elsewhere in Ballarat had begun to relax. Licence hunts were generally confined to weekdays and occasionally Saturday mornings. After midday, the defence of the stockade could be relaxed until Monday morning. Father Smyth visited and was given permission by Lalor to address the Catholics. He expressed his grave concerns over the futility of their activities and his fears that much blood would be spilt if they persisted. He exhorted them to attend Mass the following morning. Many, though well-intentioned, would never attend again.

The men continued to drill, and the manufacture of pikes continued as they all looked forward to the next day, Sunday, when the diggers would meet again at the Adelphi Theatre to reorganise the Ballarat Reform League and put forward some resolutions for relieving the crisis. In the meantime, Goodenough and his fellow spies kept Rede supplied with a constant stream of bizarre reports about what was happening in the stockade so that by Saturday evening he believed that there were companies of men formed up in well-drilled national groups. He told Hotham in a letter that there were individual companies of French, Swedish and Germans as well as Irish and Vandemonians. The Americans were not mentioned.

A deputation

That night Lalor and his lieutenants decided they would send a deputation to Rede in an attempt to prevent bloodshed. They would inform him that the diggers were prepared to lay down their weapons and return to their claims if the resident commissioner would give an assurance that licence hunts would be suspended, and the men arrested on the Gravel Pits were released. After dark Carboni, George Black and Father Smyth crossed the fields towards the camp but when they came to the bridge over Yarrowee Creek only the priest was allowed go any further. He was taken in to see Rede, who returned with him to meet Carboni and Black waiting near the bridge. Carboni could not help but be amused by Rede's appearance. Dressed in his heavily braided uniform and with one hand tucked inside his tunic, he looked for all the world like Napoleon.

Adopting the same attitude as Hotham had earlier in the week, Rede informed them he would not trade with them. He claimed that the arrested men had fired on the police and troopers and would be charged with riot, a far greater offence than merely being without a licence. They would not be released, although he was prepared to offer bail for two of the arrested men. This offer was rejected. He voiced his opinion that opposition to the licences was a smokescreen to the real agenda of the diggers, which was to mount a democratic revolution. This was hotly denied and Carboni and Black reiterated their opposition to licence hunts being held at the point of a bayonet. Carboni assured Rede that he spoke for all the foreign diggers who would lay down their arms if the licence hunts could be suspended. The meeting ended with the three delegates returning empty handed.

Father Smyth tried again later that night to no avail. Rede would not tell him when and where the next licence hunt would take place so that he could warn the diggers. Disappointed and deeply concerned, he took the audacious step of writing to the governor. He told him that the diggers were not the aggressors. They had carried no arms until forced to do so to protect themselves from attack by the government forces. Father Smyth asked again for the licence hunts to be suspended until such time as a detailed examination of goldfields administration could be completed. While he could make no promises on behalf of the diggers, he felt sure they would return to their work if this assurance was made. The letter, when it arrived the next day, nearly suffered the fate of all previous appeals, but after ordering it to be filed away, Hotham decided he should answer the priest to remind him that the law had to be enforced.

Rede interpreted the fact that the deputation had come, and the priest had returned on his own, as an indication that the diggers were frightened and wanted a way out of the dangerous situation they were in. He had already sent another message, in cipher, to Hotham that he was in need of more reinforcements. Hotham dispatched the commander-in-chief of the forces in the colonies, Major General Sir Robert Nickle, and a contingent of marines from HMS *Electra* together with an additional 600 soldiers from the 12th and the 40th regiments and four pieces of field artillery. This was to supplement the 700 to 800 soldiers already in the Ballarat camp. That left Melbourne completely undefended, from within and without, and necessitated the swearing in of several hundred special constables to maintain some semblance of law and order while the troops were away. Fortunately for the diggers, Nickle's forces found the going hard, especially as they were hauling cannon and ammunition. They were still only at the halfway point at Bacchus Marsh when the massacre at Eureka took place.

A Declaration of Independence

While Lalor's council was concerned only with the abuses of the licence system and the build-up of military might in the government camp, there was a much more republican sentiment gathering force under the direction of Alfred Black, who drew up a Declaration of Independence at the Star Hotel on the afternoon of Friday, 1 December. Frederick Vern was one of the signatories. Although he was widely seen as a leader of the diggers, his refusal to work with the council and assist in drilling the men could have weakened the situation inside the stockade so as to cause it to be abandoned. It was saved by the arrival of Captain McGill, who had gone off and enlisted the assistance of his fellow Americans in Ballarat to form the 200 strong Independent Californian Rangers whom he brought into the stockade. McGill was appointed second-in-command in place of Vern, with the responsibility for tactical arrangements. He was able to instil a little order among the diggers and was responsible for putting together a roster of sentries to guard the stockade, otherwise the massacre by the military could have been greater.

As the sentries did not know every man coming in and out of the stockade it was decided to introduce a password. Whether it was McGill or someone else who chose Vinegar Hill as the password is not known, but it was the cause of some tension because of its Irish connections. Vinegar Hill had been the site of the final battle in the 1798 Uprising of the United Irishmen in Ireland and was the term used by the convicts in the 1804 rebellion in New South Wales, most of whom were Irish. Some of the diggers resented its use in Ballarat and left the stockade. Later, the choice of password would be used to accentuate the role played by the Irish in the conflict and to discount the contribution by other British subjects and foreigners in the defiance of British authority.

When McGill heard the rumours that more troops were on their way from Melbourne, he decided to take two-thirds of his Californian rifle brigade along the Melbourne road to hinder their approach and prevent them from attacking the stockade from the rear. While the idea had merit, the force never eventuated and the stockade was left dangerously under-manned and under-armed on the Saturday night. By midnight, there were only about 120 diggers inside the stockade with about 100 rifles or pistols and enough ammunition to shoot off one or two rounds each. The rest had pikes. They were not to know that Rede had decided that a surprise attack was the best means of crushing the rebellion once and for all.

Attack

Initially Captain Thomas had considered attacking the stockade on Saturday, 2 December, but once it became obvious that many of the diggers had dispersed, he decided to delay until they were collected in one spot. He apparently didn't know that the stockade was not intended to accommodate all those willing to defend their rights and that many had gone to their own homes and the public houses after dark. He did have a good idea about the layout of the fortifications the diggers had put in place and where the buildings inside it were, as the spies had passed on this information over the course of the day.

For Rede the decision to go before light on the Sunday was perfect. He wanted to prevent the next meeting of the Reform League, which was scheduled for 2pm. The plans were made in complete secrecy to prevent informants in the camp, if there were any, spreading word of the attack to the stockade. At about 2.30am, 182 mounted and foot soldiers, and 92 police accompanied by seventeen officers left the government camp. They had all been given a glass of rum and instructed to shoot any digger who offered resistance. Only those who surrendered immediately were to be spared. To consolidate the surprise, the troops did not take the direct route up through Bakery Hill to the stockade, Instead, guided by Assistant Commissioner Amos, who had responsibility for the Eureka Lead, the column left quietly from the rear of the camp and following Yarrowee Creek to Black Hill they were able to approach the stockade without being seen. Once they had crossed the creek they regrouped on a rise 300 metres from the stockade. At around 3am, they began their final approach. It was not until they were 150 metres from the stockade that the sentry spotted them and fired his rifle.

Only then did Captain Thomas give the command to sound the bugle call to signal that firing was to commence. A small group of 40 men led by Captain Wise, who would lose his life in the battle, hurried to the other side of the stockade to act as a rearguard while the police and foot soldiers fanned out to cover the sides. They were supported by mounted police and soldiers who came around them so that the stockade was almost completely surrounded and anyone trying to escape was immediately cut down.

The diggers, woken from their sleep, grabbed whatever weapons they could find but they were no match for the oncoming soldiers. Lalor, taking command, ordered the men to cover in the shepherding holes and to hold their fire until the troops were in close range, as they could not afford to waste the small amount of ammunition they had. Outnumbered by well-equipped soldiers spoiling for blood, surrender for the diggers, like escape, had become impossible. Anyone unfortunate enough to be wounded during the firing risked being bayoneted to death once the stockade walls were breached. The pikemen, who had taken up their positions near the stockade wall, ready to impale the cavalry as it broke through, found their pikes inadequate and they were mown down by horses' hooves and the sabres their riders wielded.

Lalor himself was gunned down but his comrades had the good sense to hide him under some fallen timber, otherwise he would have suffered the same fate as so many of the diggers. Fortunately or unfortunately for Carboni, his tent was outside the stockade walls. Woken by the sound of gunfire he was unable to find a way to help his comrades. However, he was able to assist many of the wounded after the fighting was over. He was arrested and released by Captain Thomas, only to be arrested again some time later, while he was helping Doctor Carr treat the injured, by the spy Henry Goodenough who had been searching for him since the battle had finished.

The death toll was terrible. Young Captain Ross did not live to see a constable called John King haul his precious flag of the Southern Cross down and carry it away. Twenty-two diggers were killed. The government troops lost three privates and Captain Wise, and a handful of soldiers were wounded. Once the last shots had been fired, carts were sent down from the camp to collect the bodies and bring the casualties to the hospital while troops scoured the countryside making arrests of all and sundry. In total, 125 diggers, some of them quite badly wounded, were driven at the point of a bayonet up to the camp to await their fate. Other soldiers ripped through the remains of the stockade, burning tents and driving mourning women and children away. Fortunately, nobody thought to lift the slabs under which a semi-conscious Peter Lalor lay. He would make his escape to Warrenheip where he had friends who could take him in.

Chapter 8

AFTERMATH

Rede was well pleased with himself when he wrote to Charles Hotham on that fateful Sunday morning to tell him that the stockade had been taken with the loss of only a few military lives, even though one of them was the very popular Captain Wise. He seemed unconcerned about the casualties among the diggers and he justified the decision he had made to launch a pre-emptive strike by presenting his version of the events of the previous few days. The diggers had acted provocatively by building a stockade, carrying out marching drills and plundering the whole township for weapons. He believed the agitators had conscripted many well-disposed diggers against their will, and had put them to work digging trenches and building fortifications. He also pointed to the theft of Assistant Commissioner Amos' horse as a sure sign that they had planned to mount a cavalry unit.

Captain Thomas and his officers also wrote reports. In general, these were sombre affairs giving details of the military moves that had been used in the taking of the stockade. They praised their men in the performance of their duty. None of them mentioned the brutality with which the dead and injured were treated once the walls had been breached and the hopelessness of the diggers' defence was realised.

Disregarding the deaths within the stockade, Rede was more interested in the 125 prisoners he had in his very cramped lock-up. Early the next morning he moved them to the storehouse, releasing eleven of the men who could prove that they had taken no part in the stockade and were simply innocent bystanders. Of the remaining 114, Raffaello Carboni was one of the most recognisable. Everybody on the goldfields seems to have known this flamboyant little Italian with his flaming red hair and his gift for languages. He continued to protest his innocence. Henry Seekamp was also arrested on the grounds that he had used his paper to spread sedition among his readers. To Rede's disappointment none of the real leaders were among the prisoners.

Martial law

Hotham had Rede's letter by noon the same day. He immediately rushed off to the government printer to have backdated placards made announcing that evil persons were inciting the diggers in Ballarat to commit offences against the civil and military authorities. The following day, he and the Executive Council declared martial law, and Hotham began receiving congratulations from the Legislative Council and the business community of

Melbourne for his swift handling of the insurgency. Two days later, when calm was reported everywhere on the Ballarat goldfields, he was forced to rescind.

In response to Rede's request for advice on how to handle his prisoners, it was decided that there should be no immediate trials and no executions were to be carried out, except under the express orders of the lieutenant governor, thus ensuring that those arrested, for a time, were to be subject to a proper trial. Nevertheless, rumours began to circulate that gallows were being erected inside the camp so that all those arrested could be tried and hanged before any enquiry into the actions of the soldiers and the police could be carried out.

While most of the diggers had returned to work by the beginning of the new week, they did so with a heavy heart knowing that several of their comrades were dead and many more faced an uncertain future. The atmosphere of foreboding was increased when notices were posted offering rewards of £200 for information about the whereabouts of Peter Lalor and George Black. In the mistaken belief that Frederick Vern was the commander-in-chief, a price of £500 was put on his head.

Lalor had remained at Warrenheip without medical attention for the rest of that terrible Sunday. In the evening, he made his way painfully back to Ballarat where he rested in the tent of a friend. When it became clear he would die if his injuries were not attended to, he was taken to Father Smyth's house where his shattered arm was amputated. He was cared for by the priest, then moved to the tents of his friends until he was well enough to be taken by dray to Geelong to the home of his fiancée Alicia Dunne.

No one came forward with any information about George Black or Frederick Vern, who had not fired a shot at the attacking military. Vern had fled and remained well hidden. Afterwards, he was more concerned with clearing himself of cowardice among his fellow diggers than avoiding arrest.

There was no reward posted for Humffray. Nor was he arrested. He had been nowhere near the stockade. That left him free to draw up a petition with some 4500 signatures in which he appealed to the lieutenant governor to grant an amnesty to the prisoners so that an inquiry into what had happened could take place without prejudice. He told the governor that such an amnesty would give him the opportunity to confirm his confidence in the people and they in him and the British Constitution. The diggers, he said, had not taken up arms against British rule, but they had been forced to defend themselves against the misrule of those who were paid to administer the law properly. He reminded the lieutenant governor that the diggers were outraged British subjects not the foreign anarchists or armed ruffians that a deputation of leading citizens of Melbourne had been led to believe.

There were other petitions asking for an amnesty for the prisoners, all ignored by Hotham. He even rejected the advice of the chairman of the commission on the goldfields he had appointed soon after the storming of Eureka had taken place. William Westgarth was one of the most astute members of the Legislative Council. He believed an amnesty would do much to calm the situation while he and his fellow commissioners went about the business of taking evidence, not just in Ballarat, but on all the major goldfields. Hotham did, however, accept Westgarth's recommendation that he rescind the order he had given Rede to recommence the licence hunts. All three commissioners recognised that such an action could spark off a rebellion on several fields at once.

The tide turns

Rumours of angry diggers marching on Melbourne did nothing to calm the tension which had begun to brew in the city as soon as word of the storming of the stockade had become known. Meetings held for the purpose of allaying their concerns backfired as the people who came to listen to the mayor and other government officials became indignant at the explanations they were being given. At one meeting an angry throng of about 4000 people labelled the lieutenant governor despotic, and the troops murderous bullies. Some people even went so far as to suggest that the population might need to take up arms to protect themselves from the outrages of the government. The presence of police, gaol wardens and mounted volunteers in the streets around the gatherings did nothing to calm the mood of the public.

In his dispatches to London explaining what had happened, Hotham blamed anarchists and foreigners for the unrest. He was able to use a broadsheet which found its way on to the streets calling for a 'new Constitution for Victoria' based on the French and American constitutions, coupled together with Chartist principles, as a justification for the actions of the troops. This, he believed, was proof that the stockade and the agenda of the Ballarat Reform League was really about the overthrow of the British government in the colony. As a result, the arrested diggers who were brought to trial would face charges of high treason. Hotham believed they deserved the ultimate penalty. But in this, unlike all his other dealings with the diggers, he could not act alone. He had to rely on the due process of the law and trial by jury.

By January 1855, he had convinced himself of the existence of secret societies that were plotting to overthrow the government. He wrote to the Colonial Office asking for additional funds to counteract the French Red Republicans, the German Political Metaphysicians, the American Lone Star members and the British Chartists whom he thought were using the goldfields to agitate for the overthrow of the government. Surprisingly, he made no mention of Irish secret societies, which had found their way to Australia with convicts, exiles or free immigrants, and which had abounded in Ireland as an irritant to British government for the previous 100 years. Eureka must have reminded many past members of those societies of the despotic rule Ireland had endured from the British for 700 years.

The trials

At court hearings in the government camp at Ballarat, before witnesses most of the accused had never seen before, the number of prisoners held for trial in Melbourne was whittled down to thirteen; Carboni was one of them. When they arrived in Melbourne on 12 December, they were taken to Melbourne Gaol, which was under the command of the infamous John Price who had been appointed inspector of Victorian prisons following the winding down of the Norfolk Island convict settlement; he had a terrible reputation for brutality and sadism. He exercised some of this brutality on the Eureka captives by housing them four to a small cell for up to thirteen hours at a time, providing substandard rations and having them stripped naked and searched. Carboni was most affronted by this treatment, as were his companions. The prisoners were also refused writing paper during their long wait until the trials began.

The first trial was that of Henry Seekamp, and here the government scored a victory of sorts. There was considerable doubt placed before the jury as to whether Seekamp had in fact written the articles that had appeared in his newspaper. He was still found guilty of sedition and was convicted on 23 January 1855; after many appeals he was sentenced to six months' imprisonment on 23 March. He was released three months early from prison.

The trials of the Eureka captives began on 22 February 1855. William Stawell, as attorney-general, represented the Crown with Chief Justice à Beckett on the bench. Their first task was to find a jury out of the list of jurors available to them who they considered would not be biased towards the defendants. They ruled out all Irishmen and publicans and any other people they considered questionable, while the council for the first defendant, the black American John Joseph, ruled out all gentlemen and merchants. Joseph's council, the eloquent Irishman Richard Ireland, made light of suggestions that a simple black man could not have plotted anything so complicated as an attack on Her Majesty, the Queen. Joseph was acquitted by the jury to thunderous applause of the gallery.

The second man tried was John Manning, the newspaper writer. He, too, was acquitted. Stawell obtained a month's stay in the proceedings so that a new list of jurors could be drawn up. This fired up public sentiment, as it became obvious that the government intended to make a scapegoat out of those arrested and was determined to get convictions at any cost. Hotham would not listen to those who tried to show him that the justice system was being held to ridicule. He blamed the jurors for not doing their duties.

The attempt failed as defendant after defendant was acquitted by juries drawn from small business and tradesmen. By the time the trials started again on 19 March, Hotham had become so discredited that many were calling for his resignation. No jury was prepared to accept the mixture of perjured evidence and falsehoods being offered up from goldfields spies and others who could not possibly have witnessed the events. The defence presented credible witnesses such as Father Smyth and several others who could give a good account of what had actually happened at Eureka. The doctors who had attended the injured and dying on the battlefield were there to testify that Carboni had arrived late on the scene and had been arrested while helping them. All thirteen men were acquitted, leaving Hotham and Stawell as the villains in the eyes of the people, and the goldfields administration guilty of gross inefficiency, espionage and the deliberate provocation of the population.

The commission on the goldfields

The presentation of the commission's final report coincided with the conclusion of the trials. The commissioners recommended that the licence fee be abolished and replaced with a 'Miner's Right', which would cost the diggers £1 a year and give the holders full voting rights. An export duty of two shillings and sixpence per ounce was also recommended. Land near the goldfields was to be surveyed immediately and offered for sale at a reasonable price. The Legislative Council was to be expanded by twelve. Eight members would be elected by the diggers and four would be nominated to represent the major goldfields. The gold commissions were to be abolished and there was to be an increase in public services on the goldfields. In the commissioners' summing up, they commented that serious unrest among the population is usually a sign that the government is as much at fault as the people.

Once the report was published, Hotham had no choice but to act. On 2 April 1855, he wrote to the secretary of state admitting that the system of goldfields management he had inherited and enforced vigorously was unsatisfactory. He agreed with the report that the young officers, many of whom were no better bred or educated than the diggers themselves, had antagonised the diggers by their arrogance and reckless behaviour. He notified England that the system had to be abolished immediately but claimed no responsibility for the enormous cost incurred while maintaining it.

The Chamber of Commerce, clinging to its laissez faire principles, reluctantly accepted the main recommendation of the commission that the licence be replaced with an export tax. They realised the expediency of the new arrangements for the stability they would bring to the goldfields. Additionally, the licence system had long failed to provide the revenue that it was intended to bring to the economy of Victoria. It had created a burgeoning administration, the upper echelons of which had lived in an opulence that was never intended, and which absorbed nearly all the licence fees and fines it collected. The abolition of the licence system meant that the gold commissioners could be replaced by wardens at lower pay and conditions. Two-thirds of the goldfields administration, and half the police were no longer needed as the diggers readily took their Miner's Right of £1 a year. By July 1855, 30,000 Miner's Rights had been taken out and the number rose to 50,000 by the end of the year. The collection of the export tax of 2 shillings and sixpence provided none of the problems of smuggling that the pessimists had predicted.

Even before the ratification of the recommendations of the report an order to revoke the rewards offered for the arrest of Lalor, Black and Vern was given. Lalor was free to be nominated for the Legislative Council. His nomination was unopposed and he took his place in November 1855. He mellowed as time went on, and always insisted that he had no intention of overthrowing the government when he led the diggers at Eureka. He was not a democrat but he maintained that 'if democracy means opposition to a tyrannical press, a tyrannical people, or a tyrannical government, then I have ever been, I am still, and will ever remain a Democrat.' The other representative for Ballarat was John Basson Humffray.

On 12 June 1855, the *Gold Fields Act* was passed. It saw the introduction of local courts made up of members elected by the public to settle claims and deal with disputes. One of the first to be elected to the

Ballarat court was Raffaello Carboni who served for a short time, but he was disenchanted by the failure of the government to compensate him for the loss of his property which had been stolen while he was in gaol. He left the colony at the end of 1855; shortly after his book describing the events that had taken place in Ballarat in the spring of 1855 was published.

By this time Hotham was completely discredited. He tendered his resignation in November 1855 but before it had had time to reach England he was dead. He died from a fever on 31 December 1855. It has to be said that he worked exceptionally hard at the task of governing a colony that he had not wished to govern. He was a navy man and had accepted the post for the promotion it would bring him. He had been unable to delegate responsibility to, or trust, his advisors. And most of all he had been unable to accept that the Victorian goldfields were vastly different from the world from which he had come. Respect on the Victorian goldfields had to be earned through courage, understanding and fairness. It could not be demanded on the basis of accidents of birth or military or civilian rank. The diggers would accept nothing less than their rights as free men, even if blood had to be spilt at Eureka for this to be achieved.

BIBLIOGRAPHY

Annear, Robyn, *Nothing but Gold, The Diggers of 1852*, Text Publishing, 1999

Bate, Weston, *Lucky City*, Melbourne University Press, 2003

Carboni, Raffaello, *The Eureka Stockade, the Consequence of Some Pirates Wanting on Quarter-deck a Rebellion*, Melbourne University Press, 1963

Corfield, Justin; Wickham, Dorothy; Gervasoni, Clare, *The Eureka Encyclopaedia*, Ballarat Heritage Services, 2005

Currey, C. H., *The Irish at Eureka*, Angus & Robertson, 1954

Harvey, Jack, *Eureka Rediscovered: In Search of the Site of the Historic Stockade*, University of Ballarat, 1994

Historical Studies of Australia and New Zealand, *Eureka Supplement*, 1965.

MacFarlane, Ian, *Eureka from the Official Records*, Public Records Office of Victoria, 1998

Molony, John, *Eureka*, Penguin Books, Melbourne, 1984

Public Records Office of Victoria, *Eureka Documents*.

O'Brien, Bob, *Massacre at Eureka*, The Sovereign Hill Museums Association, 1998

O'Grady, Desmond, *Raffaello, Raffaello*, Hale & Iremonger, 1985

Serle, Geoffrey, *The Golden Age, A History of Colonial Victoria 1851–1861*, Melbourne University Press, 1977

Wickham, Dorothy, *Deaths at Eureka*, Ballarat Heritage Services, 1996

Withers, William, *History of Ballarat*, fascimile of the 1870 edition, Ballarat Heritage Services, 1999

INDEX